Collins *gem*

Knots

KU-185-504

Trevor Bounford

Many thanks to Isabella Percy for testing the knots

HarperCollins*Publishers*
Westerhill Road, Bishopbriggs, Glasgow G64 2QT

www.collins.co.uk

A Diagram book first created by Diagram Visual
Information Limited of 195 Kentish Town Road,
London NW5 2JU

First published 2001
This edition published 2005

Reprint 10 9 8 7 6 5 4 3 2 1 0

© Diagram Visual Information Limited 2001, 2005

ISBN 0-00-719010-7

Printed in Italy by Amadeus S.r.l.

Foreword

Knots have been in use for thousands of years. There is pictorial evidence of cords and knots from many of the great early civilisations. Knots are used as a means of securing and as a form of decoration. There are literally thousands of different named knots and no one can claim to know them all. Despite the development of modern securing methods – man-made clasps, grips, adhesive tapes and so on – knots are still universally employed to tie down loads, fasten packages and moor large and small boats, and for many other uses. Yet, apart from specialist users, very few people these days are familiar with more than one or two knots. In fact most people can recite half a dozen or so names of knots without knowing how to tie them!

This book contains many of the basic and most useful knots. It also includes some simple decorative knots, several of which may be elaborated into highly complex knots. Also included is a table of knots listing their application.

Following the clear illustrations and precise instructions, any user should be able to tie the knots in this book. With some practice it should be possible to learn to tie these knots quickly and efficiently. Not only will the ability to tie knots bring obvious practical advantages but it can also lead to the development of a stimulating, and very inexpensive, pastime.

Contents

Introduction

TYING KNOTS

Except for highly complex structures, tying knots, whether for practical use or for decoration, requires no specialist tools. The knots in this book need only fairly dexterous fingerwork, although for more complex knots tied in thinner cord, a pair of snipe-nosed pliers, preferably with curved ends, may be useful.

Material for tying will depend upon the application. Most available cord can be tied successfully although some man-made lines are not very compliant. For practice, bootlaces (those with a round profile made for walking boots) are very useful. They come in one and a half metre lengths and the hardened tip is very handy for threading through tight gaps. Small cord – strings, threads, etc. – is readily available. Rope can usually be bought from builders' merchants. Pre-stretched coloured cords can be had from boat chandlers. Decorative cords will be found in haberdashers and department stores.

Depending on the type of material and the use, the ends of thicker cord will need to be fixed in some way to prevent unravelling or fraying. Cords made from natural materials, such as hemp and cotton, can be whipped –

that is, bound round with thinner cord (*see page 142*).
Other man-made cords are sealed with heat, either by
special heated cutters or with a flame. Extreme care
must be taken both with cutting and sealing.

Tying knots efficiently requires practice. Follow the
diagrams carefully – in some cases, transposing the lay
or direction of the cord can result in a different, and
sometimes much less effective, or even useless, knot.
Once the general structure of the knot has been formed,
you should work it tight. This is done carefully by
pushing or pulling parts of the cord so the form remains
intact. Just tugging on the ends is not useful as it will
deform the knot and reduce or eliminate its
effectiveness. Except for a few particular knots, a well-
tied knot is easily untied without the need to cut it.

As in most activities, knot tying has its own language. In
this book the more obscure terms have been avoided
and the material used is generally referred to as cord,
rope or line.

The knots in this book have been grouped by type
rather than application: stopper knots; fixed loop knots;
hitches; bends; other useful knots; and decorative knots.
A few of their applications are shown on the following
pages but it is impossible to list the full range of uses.

KNOTS AND THEIR USES

STOPPER KNOTS
Used to prevent a cord
pulling through a hole,
such as at the end of a
string of beads, etc.

Overhand Knot

Double Overhand Knot

Treble Overhand Knot

Figure of Eight Knot

Stevedore Knot

FIXED LOOP KNOTS
Used to make a loop at
the end of a line, either
to slip over a post or to
pass the other end of the
line or a second line
through.

Overhand Loop

Bowline

Double Bowline

Portuguese Bowline

Figure of Eight Loop

HITCHES
Used to secure one
object to another, such as
in mooring a boat.

Lark's Head Hitch

Round Turn and Two
    Half Hitches

Clove Hitch

Constrictor Knot

Timber Hitch

Killick Hitch

BENDS
Used to attach two lines
together.

Fisherman's Knot

Sheet Bend

Double Sheet Bend

Figure of Eight Bend

Rigger's Bend

Bloodknot

OTHER USEFUL KNOTS
Knots with particular
applications.

Sheepshank
used to shorten a
length of rope or cord

Transom Knot
used to attach two
posts at right angles

Jug Sling
used to improvise a
carrying handle for a
jug or bottle

Sack Tie
secures the neck of a
filled sack

Common Whipping
prevents the end of a
rope or cord from
unravelling

DECORATIVE KNOTS
Knots which are both
practical and attractive.

Square Knot

True Lovers' Knot

Chinese Button Knot
used to make a flexible
and washable garment
fastening

Monkey's Fist
used for weighting the
end of a casting line or
as an end for a pull
cord

Turk's Head
used to bind round a
cylindrical object, such
as to make a drip stop
for an oar, or to make
a napkin holder

Chain Sennit
used as a temporary
shortening for rope

Plait Sennit

PARTS OF A CORD

Shown here are the terms used to describe parts of a cord to aid description. Note that the terms 'working end' and 'standing part' are used to describe parts of the rope at each stage and can be interchanged as the

1 Working end

The part of the cord in which the knot is being formed.

2 Bight

A curved part of cord which can be a wide or narrow curve and may touch at its base.

knot is developed. Although knots are frequently tied at the end of the cord, in some cases they may be tied some way along the line. Such a knot is described as being 'tied in the bight'.

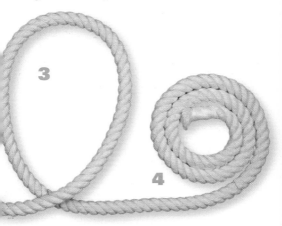

3 Loop or turn

Formed when the cord passes over itself.

4 Standing part

The rest of the cord which is not directly involved in the forming of the knot.

Stopper knots

OVERHAND KNOT
Also called Common, Simple, Single or Thumb Knot

This most basic of knots is the foundation of many more complex knots. It also has uses in its own right. With practice it can be tied in thread or thin twine with one hand drawing the working end through the loop with the thumb – hence the name 'thumb knot'.

1 Form a simple loop.

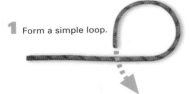

2 Pass the working end behind and through the loop.

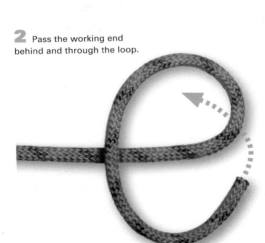

3 Pull the working end through the loop.

4 Begin working the knot tight.

5 As you work the knot tight, adjust its position to leave the required amount of working end.

DOUBLE AND TREBLE OVERHAND KNOT

These knots, also known as blood knots – possibly from use in the Cat-o'-Nine-Tails whips – are simple extensions of the Overhand Knot. Although bigger than the Overhand Knot, they have the same diameter and therefore, used as a stopper knot, will not stop a bigger hole.

1 Start with a basic Overhand Knot but, before tightening it, form a second bight in the working end.

2 Thread the working end through the main loop from the back, taking it above the first turn.

3 Pull the working end through the main loop.

4 Work the knot tight to finish the Double Overhand Knot.

5 The Treble Overhand Knot also starts as a simple Overhand Knot. It then has two additional turns. Start with stage three of the double overhand knot then form another bight in the working end.

6 Thread the working end through the centre of the loops from the back.

7 Keep pulling the working end through the main loop and begin to position the knot.

8 Having pulled the working end through, work the knot tight to complete the Treble Overhand Knot.

FIGURE OF EIGHT KNOT
Also called Flemish Knot

This is another useful stopper knot and is also a variation on the Overhand Knot.

1 Start by forming a loop, with the working end passing under the standing part. Then make a bight in the working end.

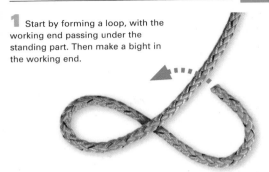

2 Pass the working end across in front of the standing part then thread it through the loop from the back.

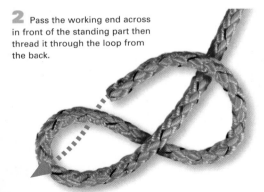

3 The knot should now have the figure of eight shape which gives it its name.

4 Work the knot tight to complete it.

STEVEDORE KNOT

A bigger stopper knot than the overhand knots. Used by dockworkers to prevent the end of a rope pulling through an eye. Basically a Figure of Eight Knot with an extra full turn.

1 Make a long bight in the working end then twist the bight to form an eye loop.

2 Put a twist in the loop.

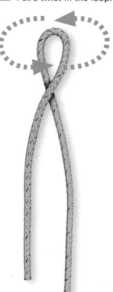

3 Twist the eye loop again.

4 Now take the working end under the standing part.

5 Keeping the twists and the loop in place, bring the working end towards the eye loop.

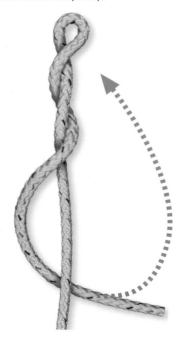

6 Pass the working end through the eye loop from the front and begin to work the twists of the knot up towards the loop.

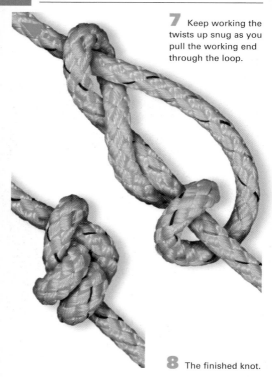

7 Keep working the twists up snug as you pull the working end through the loop.

8 The finished knot.

Fixed loop knots

OVERHAND LOOP
Also called Openhand Eye Knot and Binder's Knot

A simple loop knot made by tying an Overhand Knot in a bight. Useful when tying parcels with string or thin cord but for not much else as the knot jams tight and will generally need to be cut off.

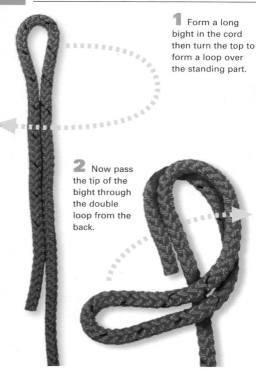

1 Form a long bight in the cord then turn the top to form a loop over the standing part.

2 Now pass the tip of the bight through the double loop from the back.

3 Keep the adjacent cords parallel as you form the eye of the knot.

4 Work the knot tight, leaving an appropriately sized eye loop.

BOWLINE

Probably the most useful fixed loop knot, the Bowline (pronounced 'bo-lin') is easy to make and is very reliable as it does not easily slip or jam tight. This is the basic version from which very many variations are formed.

1 Make a loop, with the working end passing over the standing part, and turn the end back towards the loop.

2 Pass the working end through the loop from the back.

3 Pull the working end through the loop and turn it behind the standing part.

4 Bring the end round and back to the front of the loop.

5 Thread the end through the front of the loop.

6 With the working end through the loop, begin to work the knot tight.

7 Leave a suitable size of working loop. The Bowline will hold tight but is easily loosened by forcing up the bight curved around the standing part.

DOUBLE BOWLINE

This modification makes the Bowline even more secure and is recommended for uses where the knot may get wet or be subjected to rough use.

1 Make a loop then bring the working end round to form a second loop.

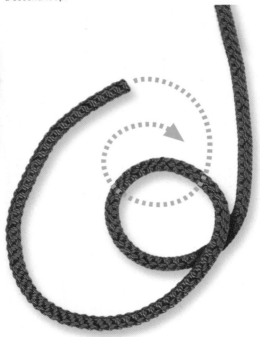

2 Turn the working end round again to pass through the two loops from behind.

3 Pull the working end through the loops and up to lie next to the standing part.

4 Take the working end round behind the standing part.

5 Thread the working end back through the front of the two loops.

6 Work the knot tight so the double loops grip the bight in the working end.

PORTUGUESE BOWLINE
Also called Spanish Bowline

The Portuguese Bowline results in a secure double loop which, among other uses, can support a person seated in one loop with the other around the chest and under the arms.

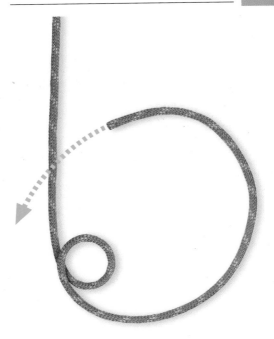

1 Make a small loop in the working end then bring the working end round to form a second, larger loop.

2 Form another large bight and turn the working end towards the small loop.

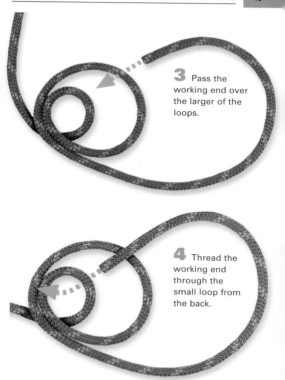

3 Pass the working end over the larger of the loops.

4 Thread the working end through the small loop from the back.

5 Pull the working end up through the small loop and over the larger loop to lie beside the standing part.

6 Pull enough line through then pass the working end behind the standing part.

7 Thread the working end back to pass over the large loops and through the small loop.

8 Work the knot tight, evening up the large loops, so that the small loop grips the working end.

9 The finished knot has two large working loops.

FIGURE OF EIGHT LOOP

This fixed loop knot can be tied in the bight if
something is to be threaded through the loop, or it can
be formed through the item to which it is to be
attached.

1 To tie the knot in the bight, make a long, narrow bight, the tip of which will form the eye of the loop knot.

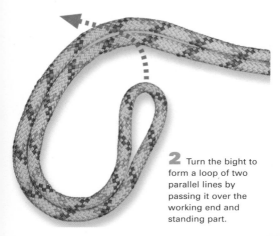

2 Turn the bight to form a loop of two parallel lines by passing it over the working end and standing part.

3 Pass the eye loop round behind the standing and working parts to make the second double loop of the figure of eight.

4 Thread the eye through the first loop from front to back.

5 Pull the tip of the eye loop through the upper loop and start to tighten the knot.

6 Work the knot tight, leaving the eye at the desired size.

7 To tie the knot onto a closed fixing, start with a loose figure of eight knot and thread the working end through the fixing.

8 Bring the working end back through the first eye of the knot, running it parallel to the existing line.

9 Take the working end around the standing part and bring it back to pass through the eye again.

10 The end should pass under the double line then back over the two that form the eye of the loop.

11 Finally, pass the working end back under the double lines to lie parallel to the standing part.

12 Work the knot tight, keeping all the lines parallel for a neat finish.

Hitches

LARK'S HEAD HITCH
Also called Ring Hitch

A very simple hitch but one that will only hold if equal pressure is being exerted on both parts of the rope. It is useful for suspending items from a ring or from a horizontal pole or line, as well as for attaching labels to parcels or luggage.

1 Pass a loop through the ring to which the hitch is being made. In this case the loop becomes the working end of the cord.

2 Pass the loop right through then open it and pull it over and around the circumference of the ring.

3 Pull the tip of the loop back to lie over the parallel standing parts.

4 Once the ring has been encompassed start to tighten the loop.

5 Draw the loop tight to secure the hitch.

ROUND TURN AND TWO HALF HITCHES

An extremely useful hitch, simple and quick to do but very secure. Used widely for mooring craft or for securing items to rings, bars, rails and posts.

1 Loop the working end around the fixing.

2 Bring the working end over the fixing again, to complete the round turn, and pass it across the standing part.

3 Take the working end round behind the standing part.

4 Thread the tip of the working end between itself and the standing part to form the first half hitch.

5 Make the first half hitch taut then take a second turn around the standing part.

6 Thread the tip through the loop formed by the working end and the standing part to make the second half hitch.

7 Work the two half hitches tight and snug up against the round turn.

CLOVE HITCH

This is a very simple hitch, basically a crossing knot formation. Although it is quick to execute, it can be jerked loose and can jam, so should be used only for securing light items and only as a temporary knot.

1 Take the working end around the fixing one full turn, crossing over the standing part where it lies against the fixing.

2 Continue around the fixing one half turn, keeping the crossover in place.

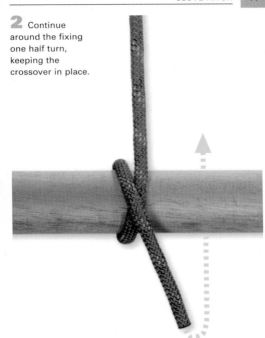

3 Turn the working end towards the crossover.

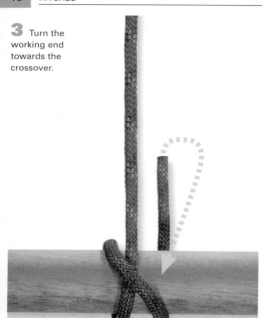

4 Thread the tip through the loop on the fixing.

5 Work the hitch tight, with the standing part and working end snug against each other.

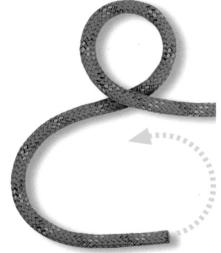

1a You can preform the knot then slip it over the end of a post. The hitch can also be made in the bight, rather than the end, if it is to drop over a post in this way. Make a loop a little way down the working end then form a second, opposing loop.

2a You should now have a figure of eight shape. Fold the upper loop over to lie on the lower loop, so that the last part of the working end lies beside the standing part. Make sure the loops are wide enough to slip over the post.

3a Slide the double loop down over the end of the post, keeping the loops and the working end together.

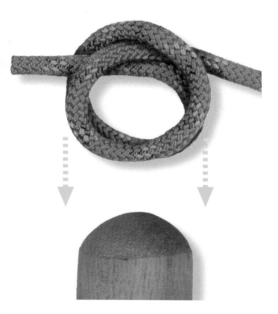

4a Slide the hitch well onto the post and work it tight, with the standing part and working end snug against each other and both trapped by the diagonal crossover.

CONSTRICTOR KNOT

The Constrictor Knot makes a very secure hitch. It is basically a clove hitch with the end secured by a trapping hitch.

1 Start by tying a basic clove hitch.

2 Take a full turn around the post, being sure to lay the working end across the standing part.

3 Thread the working end through under the diagonal but over the standing part. Then bring the working end back across, this time to pass under the standing part.

4 Work the hitch tight so that all the turns sit snug against each other and with the tip of the working end trapped beneath the loop nearest the standing part.

TIMBER HITCH AND KILLICK HITCH

The Timber Hitch is used, as the name suggests, for dragging timber. It is a simple hitch formed by a few turns in the working end made around the standing part and is only secure while under tension. The Killick Hitch is a Timber Hitch with the addition of a half hitch placed a little further along the load. This extra hitch serves to keep the log in line as it is pulled along. Both these hitches are easily loosened once they have served their purpose.

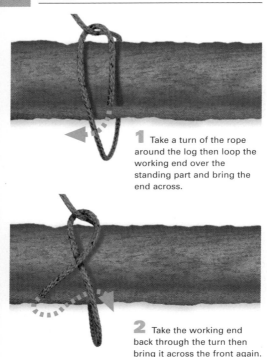

1 Take a turn of the rope around the log then loop the working end over the standing part and bring the end across.

2 Take the working end back through the turn then bring it across the front again.

3 Repeat to form a further turn around and through the main turn.

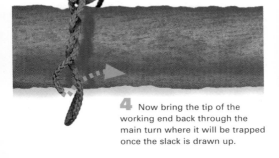

4 Now bring the tip of the working end back through the main turn where it will be trapped once the slack is drawn up.

5 Keep the eye loop and turns tight while drawing up the slack to complete the Timber Hitch. The standing part now becomes the working end.

6 The Killick Hitch is a continuation of the Timber Hitch. Take the working end one turn around the log about halfway between the Timber Hitch and the end of the log.

7 Pass the tip of the working end through under the standing part to form a half hitch.

8 Pull the rope taut between the half hitch and the Timber Hitch, making sure the loop and turns remain in place and the end is securely trapped. Take up the slack in the working end to secure the full Killick Hitch.

Bends

FISHERMAN'S KNOT

Despite the name, the Fisherman's Knot is in fact a bend, used for joining two lengths of line. It is a very simple but effective bend for joining lines of equal diameter.

1 Lay the working ends of the two lines side by side but so they point in opposite directions. With the working end of one line, start the first of the Overhand Knots by taking a turn around the second line.

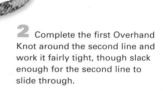

2 Complete the first Overhand Knot around the second line and work it fairly tight, though slack enough for the second line to slide through.

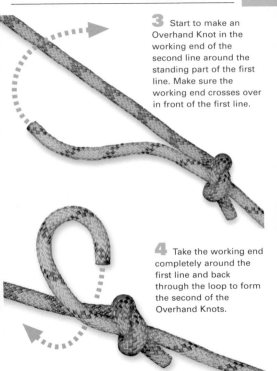

3 Start to make an Overhand Knot in the working end of the second line around the standing part of the first line. Make sure the working end crosses over in front of the first line.

4 Take the working end completely around the first line and back through the loop to form the second of the Overhand Knots.

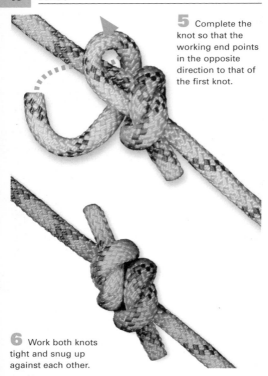

5 Complete the knot so that the working end points in the opposite direction to that of the first knot.

6 Work both knots tight and snug up against each other.

SHEET BEND AND DOUBLE SHEET BEND

The Sheet Bend, although not particularly strong, is one of the oldest known knots. It can be used to join lines of equal or unequal thickness, in which case the bight should be made in the thicker line. The Double Sheet Bend holds the bight closed more securely.

1 Form a bight in the working end of one line. Pass the working end of the second line through it then take this end round behind the bight.

2 Tuck the working end through between the bight and the standing part of the second line.

3 Work the loop tight up against the tip of the bight to complete the Sheet Bend.

4 To make the Double Sheet Bend, take a second turn around the bight.

5 Tuck the working end between the loop and the tip of the bight.

6 Work the loops tight up against the bight to secure the Double Sheet Bend.

FIGURE OF EIGHT BEND
Also called Flemish Bend

This bend is both neat in appearance and strong. Although it may jam tight in natural fibre rope, it works very well in modern synthetic cords.

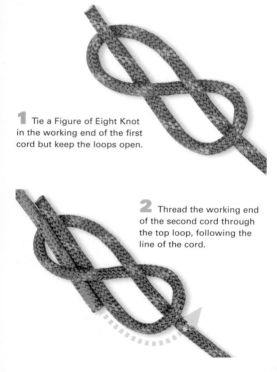

1 Tie a Figure of Eight Knot in the working end of the first cord but keep the loops open.

2 Thread the working end of the second cord through the top loop, following the line of the cord.

3 Keep threading the working end through, parallel to the existing cord.

4 Ensure that the working end passes through the loops exactly as for the first knot.

5 Keep the knot flat as more working end is pushed through.

6 Pull the working end through the final loop and work through enough cord to match the working end of the first line.

7 Now work the bend tight, ensuring that all lines remain parallel and the ends match for neatness.

RIGGER'S BEND
Also called Hunter's Bend

This is a relatively recent bend devised in the USA in the mid-20th century. It was initially publicized as a climber's bend and is very effective and neat.

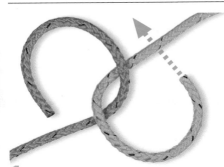

1 Make interlocking bights in the working ends of the two lines. Bring the end of one round to pass under its standing part.

2 Continue with this end round beneath the loop of the second line.

3 Thread the working end through its own loop to form an Overhand Knot trapping the second line.

4 Without tightening the first knot, follow the same pattern with the working end of the second line.

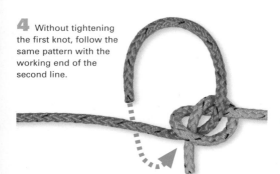

5 Take the tip of the working end through the loop of the first knot and out through its own loop.

6 Even up the ends while the knots are still loose.

7 Now work the knot tight to complete the bend.

BLOODKNOT

The Bloodknot is one of many sport fishing knots. It is used to join together two lines of equal or unequal thickness. Modern sport fishing methods employ special clamps for the purpose but traditional techniques, although more time-consuming and fiddly, are still popular among enthusiasts. In order to facilitate knots tied in synthetic line they need to be wetted, with water or saliva, before tightening.

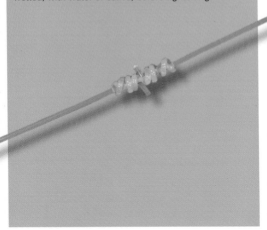

1 Work with one line at a time. Pass the working end of one line over the other line then make a turn around the second line. With fishing line this is very fiddly and requires great patience.

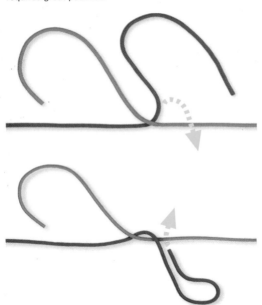

2 Continue to make more turns. The number of turns varies but four full turns should suffice.

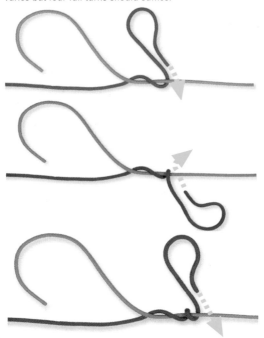

3 Keep a secure hold on both lines as you make turns.

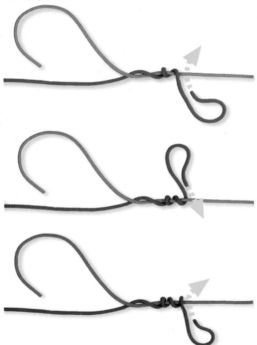

4 Once the required number of turns has been made, bring the working end back and pass it between its standing part and the working end of the second line.

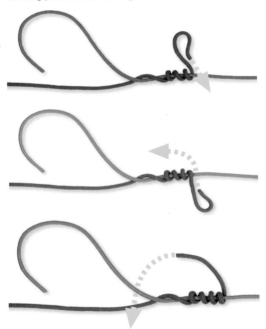

5 Now make the same number of turns with the working end of the second line around the first line.

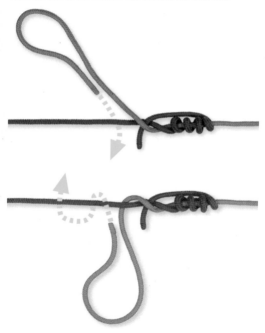

6 Ensure that the turns and the working end in the first line are held in place as you work on the second group.

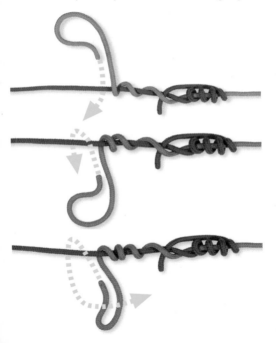

7 Having made an equal number of turns in the second line, bring the working end across and up through the central gap from the opposite side to the first working end.

8 If using nylon line you should now wet the knot. Gripping both working ends tightly (with fishing line use your teeth), pull the two standing ends on either side of the knot to tighten it.

9 You may need to twist the lines in opposite directions to pull the coils in snug. When the knot is good and tight, trim the working ends close.

Other useful knots

SHEEPSHANK

The Sheepshank is another basic knot with many variations and extensions. It can be used to shorten a rope without cutting it or to isolate and bypass a weak section of line.

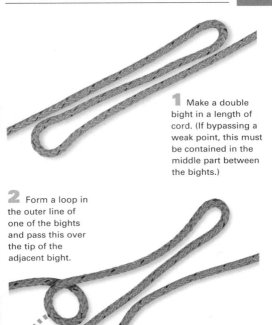

1 Make a double bight in a length of cord. (If bypassing a weak point, this must be contained in the middle part between the bights.)

2 Form a loop in the outer line of one of the bights and pass this over the tip of the adjacent bight.

3 Work the loop down so the tip of the bight extends well beyond the loop. Now form a similar loop in the other part.

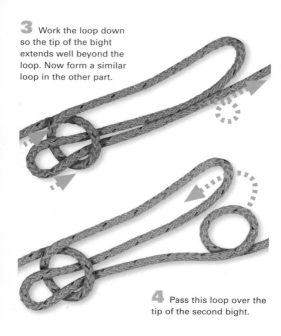

4 Pass this loop over the tip of the second bight.

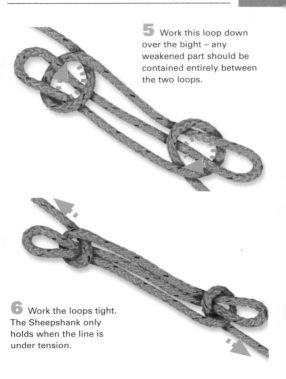

5 Work this loop down over the bight – any weakened part should be contained entirely between the two loops.

6 Work the loops tight. The Sheepshank only holds when the line is under tension.

TRANSOM KNOT

The Transom Knot is used to secure two crossed poles placed at right angles to each other. Once tied, the cord can be trimmed close to the knot or the line can extend to adjacent knots.

1 Holding the two poles together, pass a line diagonally across the junction and around behind the rearmost pole.

2 Bring the working end around to the front and lay it diagonally across the standing part.

3 Take the working end around behind the rear pole.

4 Pass the tip of the working end diagonally over the standing part then under the upper cross loop.

5 Now thread the tip of the working end under the standing part.

6 Pull both ends tight to secure the knot. The standing part can be trimmed close or may extend to form an adjacent knot (or knots) if more poles are being secured.

JUG SLING
Also called Jar Sling and Bottle Sling

The Jar Sling is another long-established knot. It was devised for the specific purpose of hanging or carrying jugs, jars or bottles. The knot will hold tight around the neck of the vessel as long as there is a lip against which it can lie. The extended loop is used as a handle.

1 Form a broad bight in the cord then turn it back over the standing part to form a double loop.

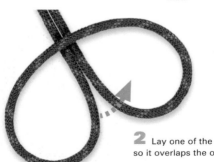

2 Lay one of the loops so it overlaps the other.

3 Draw the central point of the original bight back between the two standing parts and down into the rearmost loop.

4 Take the tip of the working bight over the adjoining loop and down through the aperture formed by the overlapped loops.

5 Pull the tip fully through the centre and up through the second of the overlapped loops. Turn the knot over.

6 Pull the bight nearest the tip of the loop back to lie at the top of the knot. Turn the knot over.

7 Once again pull the bight nearest the tip of the loop back to lie at the top of the knot.

8 The formed knot can be slipped over the neck of a vessel and tightened by pulling the loop.

SACK TIE WITH DRAW LOOP

Despite the widespread use of plastic packaging, the Sack Tie is still a useful knot with the specific purpose of securing the neck of a fabric sack. The addition of the draw loop makes untying much easier, although the knot may be tied without one.

1 Make a full turn around the closed neck of the sack, crossing the working end over the standing part.

2 Pass the working end around once more and form a bight in the end.

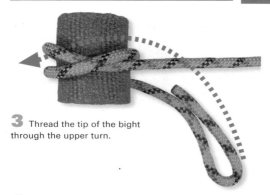

3 Thread the tip of the bight through the upper turn.

4 Pull on the bight and the standing part to tighten the knot. To untie, pull the tip of the working end to draw the bight out of the round turn.

COMMON WHIPPING
Also called Seizing

Whipping is used to bind or seize the end of a rope to prevent it unravelling. Common Whipping is the simplest form. Waxed thread is easiest to work with and will form a secure binding. Sticky tape will hold the end temporarily.

1 Lay a long bight along the rope with its tip at the end. Make a full turn round rope and bight.

2 Make 8 to 12 more tight turns.

3 Thread the working end through the eye.

4 Pull the eye back within the binding.

5 Trim the ends close to the binding.

6 Remove the sticky tape.

Decorative knots

SQUARE KNOT
Also called Japanese Crown or Japanese Success Knot

This is a simple symmetrical knot that works well as a
fastening for a neckscarf. In North America, Square
Knot is the name given to the Reef Knot and the
two should not be confused.

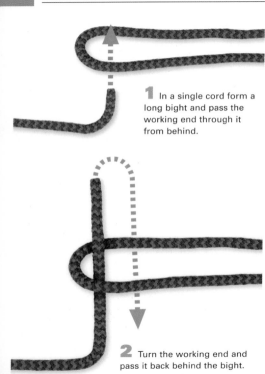

1 In a single cord form a long bight and pass the working end through it from behind.

2 Turn the working end and pass it back behind the bight.

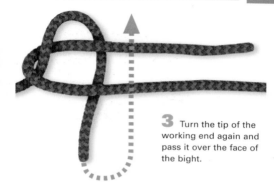

3 Turn the tip of the working end again and pass it over the face of the bight.

4 The first bight now becomes the working part. Pass the working end of this over the first bight of the standing part and through the second bight.

5 This completes the formation of the knot.

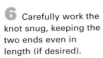

6 Carefully work the knot snug, keeping the two ends even in length (if desired).

TRUE LOVERS' KNOT

The True Lovers' Knot is formed by interlocking two Overhand Knots in a single length of cord and drawing out a bight on each side.

1 Make a loose Overhand Knot part way along the cord. Then thread the working end through the main loop from front to back.

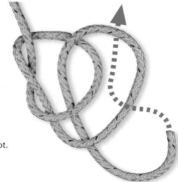

2 Pass the working end down behind the standing part and form the second Overhand Knot.

3 Draw the loop of one of the Overhand Knots through the crossover of the opposite knot, keeping the bight between the two knots intact.

4 Similarly, draw the loop of the second knot through the opposite crossover.

5 Even out all three bights while working the rest of the knot tight.

CHINESE BUTTON KNOT

Button knots make a decorative, and durable, fastening for garments. The Chinese Button Knot is a simple, traditional design and can be made with plain or decorative cord. The button knot is attached to a garment by stitching down the two parallel cords. A retaining loop of the same type of cord is fixed to the opposite flap of the garment so that the button passes through, and is then held by, the loop.

1 Start the knot by forming two opposing loops then partially overlap the loops so the rearmost loop lies over the foremost.

2 Bring the working end from the right-hand loop across to form a bight.

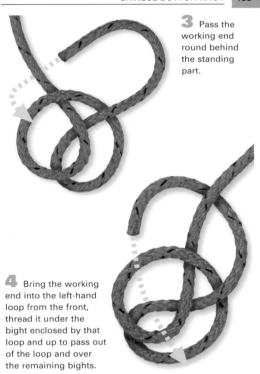

3 Pass the working end round behind the standing part.

4 Bring the working end into the left-hand loop from the front, thread it under the bight enclosed by that loop and up to pass out of the loop and over the remaining bights.

5 Turn the working end back and pass it over the outer loop, down under the two crossing bights and up out of the rectangular shape. The other end now becomes the working end.

6 Turn the knot round and pass the new working end parallel to the other end through the same rectangular hole, but keeping the loop between them.

7 Start to work the knot tight. The ends will need to be trimmed to equal length when the knot is complete.

8 You will need to work the knot backwards and forwards to tighten it without distorting the shape, which should be almost symmetrical and in the form of a knobble.

MONKEY'S FIST

This is both a practical and a decorative knot. It was originally used to weight the end of a line to be thrown from ship to wharf, or vice versa, and as such would be formed around a pebble or some other suitable object. Made from fancy cord, it serves as a handle for a curtain pull or simply as a decorative finish. The name is derived from the knot's resemblance to a small, clenched fist gripping an object.

1 Form three equal full turns in a length of cord.

2 Lay the turns exactly over each other to form a coil.

3 Turn the working end 90° and start to wind the cord around the existing coil.

4 Make three full turns around the first coils while keeping the original turns together and without distorting the shape.

5 Now thread the working end under the second coil and out at the base of the knot above the first coil.

6 Pull the working end down close to the bight that connects the first and second coils. Turn the knot through 90°.

7 Thread the working end through the gap between the top of the second coil and the underside of the first coil, then around to pass down again.

8 Thread the working end back through the similar gap at the base of the knot, keeping the turn close against the outer coil. Turn the end back up again.

9 Take the working end again through the gap between the first two coils.

10 At this stage, before all the coils are complete, you can add a marble, pebble or any other spherical object of suitable size. This will add weight to the finished knot and help maintain its shape. Don't insert anything too heavy, such as lead or any other metal, if the knot is intended for a casting line as this could cause injury to an unsuspecting catcher.

11 Now make the final turn up through between the two groups of three turns, keeping the shape of the knot intact.

12 Complete the final turn by threading the working end through and out of the base of the knot on the opposite side to the standing part.

13 Work the knot tight by pulling each part of the cord from one end through the knot to the other end. Then work back through the knot, keeping its shape until it is completely snug.

TURK'S HEAD

The Turk's Head is another decorative knot that originally had a practical use. In this case it was tied around oars as a drip stop or to prevent slipping. It will serve as a grip on a tool handle or a rail. The simple form of the knot serves as a 'woggle' to hold a neckerchief. More complex forms can be developed from this basic structure.

1 Make the Turk's Head in situ if it is to be a handle or grip around an item. To make a woggle or napkin holder use a suitably sized cylindrical object such as a cardboard mailing tube. Tape one end of the cord to the base then take a full turn round, crossing the standing part once. Take a second turn round the base and bring the working end up under the first turn and out to the left side.

2 Now rotate the base about 90°, bringing the working end down towards you.

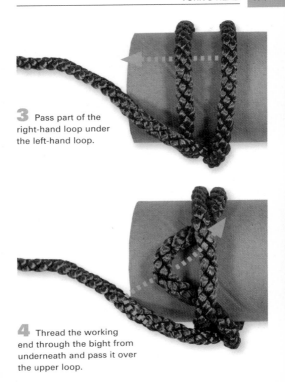

3 Pass part of the right-hand loop under the left-hand loop.

4 Thread the working end through the bight from underneath and pass it over the upper loop.

5 Rotate the base again, as in step 2, then thread the working end under the right-hand loop and up between the two loops.

6 The knot has now made a complete circuit but with one interwoven strand. Start the second circuit by passing the working end up through the loop so it lies adjacent to the standing part.

7 Continue to thread the working end following the first lead.

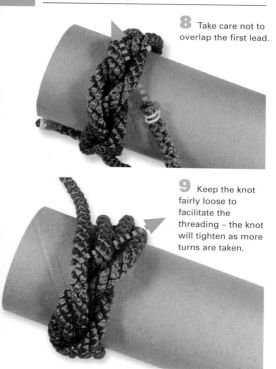

8 Take care not to overlap the first lead.

9 Keep the knot fairly loose to facilitate the threading – the knot will tighten as more turns are taken.

10 Having completed the second circuit, start the third in the same way, threading the working end through parallel and adjacent to the first two.

11 Once the third circuit is complete the knot needs to be worked tight. Work backwards through the knot, piece by piece, drawing out any slack. On working right through to the standing part it may be necessary to go back again from that point. When the knot is fully tightened it can be finished by trimming the excess from the working end and standing parts and tucking these inside the knot.

If the knot is to be removed, take it off carefully and finish it by trimming the ends inside. The ends of a loose woven cord may need to be stitched or glued together.

CHAIN SENNIT
Also called Chain Sinnet

The basic Chain Sennit or Sinnet is a form of knotting used to take up slack rope in a way which can be quickly released. In this form it is frequently seen in the circus big top and in theatres. It is also used decoratively with many elaborations based on its form.

1 Make a loop near the top of the cord.

2 Add a bight below the loop. Pass the bight through the loop from back to front to make a hanging loop.

3 Make a second bight and pass it through the hanging loop, also from back to front, to form another loop.

4 Keep adding bights and loops to the chain until all the slack has been accommodated.

5 After forming the last loop, the working end is threaded through it. By taking the working end out of this loop and pulling on it the chain is immediately dismantled.

PLAIT SENNIT
Also called Plat, Pleat or Braid Sennit or Sinnet

Plait Sennits vary from the simple three-ply braid, often used for plaiting hair, through to very complex decorative work of many strands. Shown below are a Three-ply Flat Sennit, a Four-ply Flat Sennit and a Four-ply Square Sennit.

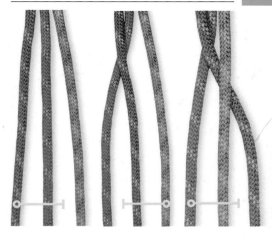

1 Start the Three-ply Flat Sennit by laying the cords side by side. Pass the left (purple) cord over the middle (blue) cord. The blue becomes the left cord and the purple is middle.

2 Now pass the right cord over the middle cord.

3 Pass the left cord over the middle cord.

Note that in order to make the sequence as clear as possible, three different coloured cords are shown, whereas normally the cord would be of one colour.

4 Pass the right cord over the middle cord.

5 Left cord over the middle cord.

6 Right over middle.

7 Continue with the sequence, working the plait tight from time to time, until the desired length is reached. The ends can be secured with a Constrictor Knot (*see page 85*).

1 For a Four-ply Flat Sennit lay the cords side by side then pass left over left middle and right middle over right.

2 Pass right middle over left middle.

3 Right middle over right.

Note that different coloured cords have been used here for clarity but cords of a single colour or matching pairs can be used.

4 Left over left middle.

5 Right middle over left middle.

6 Right middle over right.

7 Left over left middle.

8 Right middle over left middle.

9 Right middle over right.

10 Left over left middle.

11 Right middle over left middle to finish one cycle.

12 Work the plait tight then repeat steps 1 to 12 as required.

1 The Four-ply Square Sennit is usually worked with two colours of cord but four separate colours are used here for clarity. Secure the cords at the top then lay them side by side. Start by passing the left cord under the left and right middle cords then back over the left middle cord. (Note that the cords adopt the name of their new positions.)

2 Bring the right cord under the right and left middle cords.

3 Left middle over right middle.

4 Left under left middle and right middle.

5 Right middle over left middle.

6 Right under right middle and left middle.

7 Left middle over right middle.

8 Left under left middle and right middle.

9 Right middle over left middle.

Continue with the sequence, working the plait tight as you go. When the desired length is achieved, the ends can be secured with thin cord.

Index